**NATIONAL GEOGRAPHIC**

School Publishing

# Animals of Denali

## PATHFINDER EDITION

By Susan E. Goodman

## CONTENTS

# Animal

# De

*Denali National Park in Alaska is a harsh land. It's too cold for most people. But for many plants and animals, Denali is a paradise.*

A brown bear and her cub

*...of Denali*

By Susan E. Goodman

Red fox

Caribou

Great horned owl

3

A grizzly bear stumbles out of her den. She is slow and clumsy after her long winter sleep.

The bear is not alone. Her cubs follow her. She gave birth to them while hibernating, or sleeping.

Now winter is finally over. The snow is melting, and the days are getting longer. The cubs are ready to play in the sunshine.

Spring has come to Alaska's Denali National Park. But life there isn't always so playful. While the bears slept, other animals struggled through a long winter.

## Harsh Winter

Denali National Park lies in central Alaska. Forests, glaciers, and snowy peaks are found there. Mount McKinley, the highest mountain in North America, is also there. Larger than Massachusetts, the park was created in 1917 to protect wildlife.

Denali's animals need lots of land. Wolves, for instance, need a vast area in which to hunt. A single pack's territory can be 389 square kilometers (150 square miles).

Much of the park is **tundra**—vast, open land with grasses, shrubs, and no trees. It's a hard place to spend the winter.

In December, the sun hangs low in the sky. There are only four or five hours of sunlight each day. That's not much time to warm the land. Temperatures can dive to 50° below zero. Deep snow blankets the tundra.

## On the Move

Few animals live on the tundra during the harsh winter. Moose and wolves brave the cold and snow. They spend the short days looking for scarce food.

Like bears, some animals **hibernate** during winter. They eat enough food in the fall to survive the long winter in their dens.

Other animals **migrate**. They travel south or head into forests, where temperatures can be warmer and food is easier to find.

Among the migrating animals are caribou, also known as reindeer. Their broad, flat hooves act like snowshoes. These hooves help caribou cross icy and soggy ground.

In spring, warmer temperatures return to Denali. So do many plants and animals.

**Pack Animal.** Gray wolves often travel in packs. Parents and pups usually form a pack. BELOW: Caribou migrate across the autumn tundra. BELOW RIGHT: Owls are predators that hunt at night. They prey on mice and other small animals.

## Spring Blossoms

As Denali's long winter changes into spring, the landscape comes alive. Colorful plants blossom. Animals shed their thick winter coats. Some even change color.

All winter, the snowshoe hare wore a white coat. White fur helped the animal blend in with the snow and escape hungry **predators**.

Now the hare sheds its winter coat. Brown fur grows in its place. The new color matches the ground, so the hare can easily hide while nibbling on grass and buds.

## Summer Visitors

Come summer, the tundra gets almost crowded. Birds fly in from as far away as Africa and China. Herds of caribou return from their winter home in the forests. Young caribou are born in late May, as their mothers make their way back to the tundra.

Minutes after birth, a caribou calf is unstable. It can walk only short distances. Yet just a day later, the calf can run faster than a human. It has to move quickly. Hungry wolves are often nearby.

## Wolf Pack

Wolves usually hunt in packs. The leader of the pack goes first. The other wolves follow. They can travel more than 48 kilometers (30 miles) a day looking for food.

When a pack sneaks up on a group of caribou, the wolves search for a young or sick animal. They rush in to separate their target from the rest of the herd. Then the hungry wolves attack.

Wolves may seem cruel. But they are just trying to survive. And their actions help other animals. Ravens, foxes, and even grizzlies feast on the predators' leftovers.

Oddly enough, wolves even help the caribou. Killing sick ones helps the rest of the herd stay healthy. If a herd grows too large, its members could eat all the food in the area. Then many caribou would starve.

## Summer Sun

In midsummer, the tundra is a land of plenty. The air is filled with birdcalls and the buzzing of insects. Flowers carpet the ground in pink, purple, and yellow.

Food is easy to find. Moose and caribou graze on plants in the tundra's many ponds. Owls swoop down from the sky to grab hares with their **talons**. Grizzlies gobble roots and blueberries.

Grizzlies also eat other animals—from squirrels to caribou. But bears would rather feast on leftovers than hunt their own food.

To look for food, a grizzly stands on its hind legs, sniffing the air for scents. A grizzly's keen sense of smell helps it find food.

## Plentiful Pests

Summer's warm temperatures also bring pesky insects, including mosquitoes. The tundra has more of these bloodsuckers than anywhere else on Earth.

Caribou can lose a lot of blood a week from mosquito bites. To escape mosquitoes, caribou sometimes curl up in patches of leftover snow. Who wouldn't?

## The Big Chill

At the end of summer, days get shorter. A chill fills the air. Mosquitoes disappear.

Other animals get ready for winter. Dall sheep begin growing thicker coats. Grizzlies start eating—a lot.

During the fall, grizzlies spend much of their time eating. They gain a lot of weight. Packing on fat keeps bears nourished in the cold months.

The tundra's summer visitors start heading to their winter homes. Herds of caribou travel back to the forests.

The arctic tern has much farther to go. This bird migrates all the way to Antarctica—more than 32,187 kilometers (20,000 miles) away.

## Deep Freeze

By November, snow has been falling for some time. Food is hard to find. Grizzlies stop eating and start searching for dens. Soon they'll crawl into their dens and sleep through the winter.

Now blanketed with snow, the tundra may seem still and lifeless. But it's not. Wolves are on patrol, looking for prey. Hunting is much harder now. Hungry packs look for moose and Dall sheep, weakened by illness or hunger. Wolf howls echo over the flat, snowy land.

A long, hard winter lies ahead. But spring will return one day. When it does, the cycle of life on Denali's tundra will start again.

*What habitats are near your community? What kinds of wildlife live in those habitats? What can you do to protect them?*

**High Point.** Wind blows snow off Mount McKinley, also known as Denali. ABOVE RIGHT: Dall sheep live on many of the peaks in Denali National Park.

# Wordwise

**hibernate:** to rest during winter

**migrate:** to move

**predator:** animal that eats other animals

**talon:** claw

**tundra:** cold, treeless plain

# The Making of a Park

**Alone in the Snow.** Denali National Park is a perfect place to be alone in nature. Yet it took many people working together to protect this wild land.

Alaska may not be the easiest place to live. The winter weather can be bitterly cold. In December, the sunlight lasts only a few hours a day. And with few roads into the wilderness, the best way to travel is often by plane, boat, snowmobile, or sled. It's no wonder the state has fewer people per square mile than any other. This is a wild land. But despite the rugged terrain, many animals still call it home. Why? It's thanks in large part to Denali National Park.

**Landing on Water.** Parts of Alaska have few roads. So people often use seaplanes to travel through the state.

9

## A Land of Parks

One thing Alaska has is land—lots of it. Alaska is twice as large as any other state.

Denali National Park takes up more than 2.4 million hectares (6 million acres) at the center of the state. The park protects many kinds of plants and animals.

Today, Denali is one of many national parks. But Denali is special. It was the first national park in Alaska. One man—and a whole country—made it happen.

## One Man's Vision

In the early 1900s, a man named Charles Sheldon had an idea. He dreamed of a place where animals could live wild and free, without being overhunted.

It would be a place for humans too. People could see wildlife in a wild land. They could experience the thrill of hiking through the rugged outdoors.

At the time, there were only a few national parks. The first was in Wyoming. It was Yellowstone National Park. Sheldon hoped to establish another in Alaska. Yet he wouldn't be able to do it without a fight.

## Struggle Over Land

Not everyone liked Sheldon's idea. Many Alaskans depended on the land for a living. The park might change everything!

Miners dug for gold, silver, copper, and lead. What would they do if they couldn't mine? Many were afraid of losing their jobs.

Other people hunted animals for their meat and skins. Where would they hunt if Sheldon got his way?

But some people believed the park was a good idea. They thought Alaska's wildlife might be lost forever without a park to protect the plants and animals.

**Hiding Out.** This snowshoe hare is one of the many animals that live in Denali all year long.

**Hiking Haven.** Denali has many trails that let hikers enjoy nature.

**Wild Land.** Denali protects a large area of land for wildlife. Caribou and other animals can roam for miles in the park.

## Turning Heads

In January 1917, NATIONAL GEOGRAPHIC magazine ran an article about the park. The article explained how much the animals needed the park. It said that without the park, they would likely be killed off. The land would forever be changed.

The article turned many heads. Suddenly, people across the country got involved. They wrote their congressmen. Congress listened. The next month, it voted to create the park. President Wilson soon agreed, and the country had its newest national park.

## Bigger and Better

Over time, the park has grown. In 1980, almost 1.6 million hectares (4 million acres) were added. Now the park is four times its original size!

Today, Denali National Park protects more than a thousand different kinds of plants and animals. Few people live near Denali. But more than 300,000 people visit the park each year.

Charles Sheldon would be happy to see Denali today. His dream came true. Denali is a place for animals—and for people too.

# Life in Denali

**It's time to discover how much you've learned about Denali.**

**1** What is winter like in Denali National Park?

**2** How do animals survive the winter weather?

**3** What changes take place in Denali as spring arrives?

**4** How does living in groups help some animals survive in Denali?

**5** What challenges did people face when creating Denali National Park?